The UAV Question and Answer Book

Predator in Flight

Author: Paul Love

Copyright 2011 Ioxbooks

Contact Information
Email address: ioxbooks@gmail.com

ioxbooks (ioxcom)

CONTENTS

Introduction

Are manned aircraft about to go the way of the Pony Express or silent movies? Has the last generation of manned fighter pilots already been born? Probably not, but it is clear that unmanned aerial systems are going to have an increasing impact both on the battlefield and on commercial air operations. UAVs or Unmanned Aerial Vehicles have already become a vital part of the United States military arsenal and they are also finding new applications in civilian life every day.

Modern military UAVs are revising the the whole concept of aerial surveillance and close air support. Providing extensive coverage of the battlefield and delivering lethal payloads with surgical precision, UAVs are guided by satellite and can be directed by personnel sitting in a trailer halfway around the world. They have proven to be a major asset in locating and targeting terrorist activities in places like Iraq, Afghanistan and Serbia and in providing vital information to field commanders and troops on the ground.

Similarly, unmanned aerial vehicles are showing up outside of the military doing jobs such as:
- Gathering data from inside a hurricane to aid in scientific research
- Performing border patrol duties for the U.S. Customs service
- Providing fire fighters with thermal imagery and mapping information on wildfires

3

- Taking part in search and rescue operations, aiding in disaster response, transporting cargo, performing crop dusting, and providing police with an "eye in the sky"

This book is concerned primarily with the use of UAVs in the United States and is intended to provide a quick overview of what UAVs are, how they're being used, how they operate and what impact they may have on the present and future.

Note: This book is dedicated to the men and women of the 32nd Tactical Reconnaissance Squadron who (many years ago) let a young Army soldier work alongside them. And to the F-4 Phantom - that big, fast, powerful, tough as nails, ugly as sin, and extremely dangerous "hunter in the sky".

UAVs In General

How did UAVs come about?

The earliest use of an unmanned aerial vehicle in war may have happened in 1849 when the Austrians used unmanned ballons loaded with explosives in an attack on the the Italian city of Venice. Later, aerial torpedoes were used in World War One - these "flying bombs" used mechanical gyroscopes to maintain straight and level flight. They were launched in the direction of enemy forces and when they ran out of fuel they would fall out of the sky and explode on impact. Later UAVs used radio guidance systems to carry out reconnaissance missions under control of a human pilot. The invention of integrated circuits eventually led to the development of today's UAVs – unmanned aircraft that can fly themselves to a large extent and can also be operated by a ground controller.

What were some of the earliest UAVs?

Probably the first unmanned aerial vehicle was A.M. Low's "Aerial Target" in 1916. That was followed by a number of remote-controlled aircraft, including the Hewitt-Sperry Automatic Airplane. Later, drones were used during WWII for target practice and jet engine powered UAVs appeared following WWII, but UAVs didn't really come into wide-spread usage in the military until the Vietnam War.

Were UAVs ever used in WWII to attack a target?

In July of 1944 a 4-drone attack was launched against a beached Japanese merchant ship in the Russell Islands. The attack resulted in two hits and two near-misses.

Where did the term "drone" come from?

The use of the word "drone" in connection with UAVs may stem from the British development in the 1930s of an unmanned radio-controlled aerial target named the "Queen Bee".

What was Operation "Red Wagon"?

After Gary Powers and the U-2 spy plane he was flying were shot down over the USSR in 1960, the U.S. started a classified UAV development program called "Red Wagon". The program resulted in the development of UAVs (Ryan reconnaissance drones) that flew over 3400 missions for the United States during the Vietnam war.

What is a "UAS"?

In many cases the term "Unmanned Aerial System" is used instead of "UAV" to emphasize that there is an entire system involved in UAV operations, not just a "Vehicle". In most cases the system includes an aircraft, one or more control stations, and a communications and control link. For example, the RQ-7 Shadow UAS consists of four unmanned aircraft, two ground control stations (GCSes), one portable GCS and one Remote Video Terminal (RVT).

What is a "UCAV"?

An armed UAV is referred to as an unmanned combat air vehicle or UCAV.

What is an "MAV"?

Micro Air Vehicles or MAVs are proving to be effective tools for soldiers and police operating in the narrow confines of city buildings or crowded areas. MAVs are limited by short flight times since they can't carry much fuel or heavy-duty batteries because of their size, but new versions with longer flight times are under development. Current MAVs include:

- XQ-138 – Developed at the University of Kansas, the XQ-138 takes off and lands on its tail. It can fly at speeds up to 150mph by using actuators to counter wind gusts and keep it stable in flight.
- Cyberquad – Built by Cyber Technology, the electric-powered Cyberquad uses four ducted fans that let it fly and hover like a helicopter. The Cyberquad was used recently to inspect a damaged oil rig in the Timor Sea.

What does "sUAS" stand for?

sUAS stands for Small Unmanned Aircraft Systems. Advances in technology are making it possible to pack the capabilities of older, larger UAVs into smaller and smaller airframes and the term SUAS refers to these smaller UAV systems.

What does "STUAS" stand for?

STUAS stands for Small Tactical Unmanned Aircraft Systems and refers to micro-UAVs.

What does "RPA" stand for?

In the last few years the U.S. Air Force has begun referring to larger UAVs like the Predator, Reaper and Global Hawk as Remotely Piloted Aircraft (RPA) to emphasize that these systems are always controlled by a human operator. The term Remotely Piloted Vehicle (RPV) has also been used.

What's the difference between a cruise missile and a UAV?

The basic difference is that the UAV is reuseable and the cruise missile is not. A secondary difference is that a UAV can be used for reconnaissance, surveillance and target illumination in addition to being used to destroy a attack and destroy a target.

What's the difference between a "drone" and a UAV?

Unmanned aircraft like the ones employed by the U.S. in Vietnam were used for reconnaissance and basically just flew in a straight line or preset circles taking imagery until they ran out of fuel. Because of their relatively simple nature they were referred to as "drones". As new models were developed that were controlled by an operator the military began to use the term "RPV" (Remotely Piloted Vehicles). The term "UAV" came into use as unmanned aircraft began to combine remote operation with onboard piloting capabilities, ranging from simple flight plan programming to systems that allow the UAV to determine its own flight path and maneuvers.

What's the difference between a UAV and a R/C (radio controlled) airplane?

To be considered a UAV, an aircraft's primary purpose needs to be military or commercial, not recreational. In general aircraft weighing less than 55 pounds aren't considered "vehicles" unless they have considerable autonomous operational capability, which means normal R/C models aren't considered UAVs.

Are military UAVs "flown" by a human pilot?

Yes and no. Most of the larger UAVs are capable of flying a pre-programmed flight plan with the human pilot taking control only if it's necessary to make some change in the route.

What are the basic types of UAVs?

There are two basic varieties of unmanned aerial vehicles – ones that have to be controlled from a remote location and ones that can be controlled manually or fly autonomously (on their own, without human guidance).

What is an "autonomous UAV"?

Autonomous UAVs are capable of carrying out certain types of flight operations on their own, such as:

- Determining their own flight path based on their programming and on-board sensor readings
- Selecting the best maneuvers to perform a specific task
- Initiating communications with ground stations and other UAVs

Where will the biggest improvements in UAVs be made in the near future?
The current trend seems to be moving away from improvements in the technology that lets pilots fly the UAV and toward greater autonomy for the UAV itself. More sophisticated electronic systems are being developed to let the UAV "think" for itself and make tactical decisions (short of launching an attack) on its own.

What are the different functional categories of UAVs?
- Reconnaissance - provide battlefield intelligence for the military
- Combat – armed UAVs that provide an attack capability in high-risk environments
- Target – provide ground and aerial gunnery targets
- Logisitics – UAVs designed for cargo carrying
- Civilian – UAVs specifically designed for civilian use
- Research and Development – used to test new designs and new technologies

What types of missions are assigned to UAVs?
Unmanned aerial vehicles fly reconnaissance or attack missions (or both). UAVs can also provide laser targeting for attack helicopters.

Are there any anti-UAV weapons?
U.S. Air Force researchers are working on a couple of different anti-UAV weapons. One approach uses high-energy bursts of microwaves to disrupt a UAV's electronics and disable the aircraft, while a second

approach uses a Humvee-mounted laser to shoot down enemy drones. Both programs reflect a growing concern over enemy forces using small unmanned aircraft to spy on or even attack U.S. Forces.

How long can a UAV stay airborne?

Longer than a manned aircraft (assuming the manned aircraft doesn't refuel in-flight). In 2006 a solar-powered Zephyr UAV stayed aloft for 54 hours and some current military UAVs can remain airborne for flights lasting close to 40 hours.

Can UAVs refuel in flight?

Not yet, but soon. On July 1st, 2010 DARPA (Defense Advanced Research Projects Agency) gave a $33 million dollar contract to Northrop Grumman to develop an aerial refueling system using two Global Hawks. According to initial press releases the system will also allow UAVs to refuel at much higher altitudes than conventional aircraft, which normally refuel somewhere between 20,000 and 35,000 feet.

Are all UAVs controlled from the ground?

Normally UAVs are handled by a controller on the ground. However, experimental flights were done several years ago to test a UAV Command and Control Interface (UAVCCI) that would allow a single manned fighter to control multiple UAVs while airborne. The UAVs are semi-autonomous, meaning that they can fly straight and level on their own and the pilot of the manned fighter only has to give them orders for specific actions such as attacking a specific target or executing a search pattern.

11

How are UAV flight plans programmed?

UAVs like the Reaper or Global Hawk are usually set up to follow a set of "waypoints" fed into the UAV's GPS (Global Positioning System) receiver. Waypoints are a set of coordinates (longitude, latitude, altitude and possibly time) that identify a particular point in space (or space and time).

Has a military UAV ever gone out of control in a civilian area?

Yes, at least on one occasion. In August 2010 Navy operators lost control of an MQ-8B Fire Scout. The operators finally regained contro but not before the UAV entered restricted airspace around Washington, D.C. The incident was apparently the result of a software problem that caused the Fire Scout to ignore its pre-programmed flight procedures.

Are there restrictions on where a UAV can fly in the United States?

The Federal Aviation Administration doesn't permit UAVs to fly in the national airspace, primarily because of the limited visibility of UAV operators. The FAA feels that the lack of full peripheral vision for the UAV pilot could lead to a midair collision with a manned aircraft. However, as of October 2010 the Army has been given tentative approval to fly UAVs in the national airspace at night using ground-based radar and GPS systems to avoid civilian and commercial air traffic. The Army plans on implementing a ground-based "sense and avoid" technology integrated into their UAS control stations which should prevent any mishaps.

What's the smallest UAV?

AeroVironment is testing its "Nano Hummingbird" UAV, which is designed to look and fly like a hummingbird. The Nano Hummingbird has a wingspan of just 6 ½ inches (16 cm), weighs just 2/3 of an ounce (19 grams), and can climb, descend, move sideways, fly forward and backward and clockwise and counterclockwise. In spite of it's tiny size, the Nano Hummingbird contains a motor, batteries, a video camera, and a communications system so it can be flown by remote control.

What's the largest UAV?

Currently AeroVironment's Global Observer is the largest UAV in operation. Global Observer measures 70 feet in length, with a wingspan of 175 feet. It can operate at 55,000 to 65,000 feet and stay aloft for up to 7 days while carrying a payload of up to 400 pounds (180 kg). Military mission capabilities include acting as a satellite-type communications system, allowing different teams to communicate over long distances. Possible civilian uses include weather monitoring, emergency communications link in case of natural disaster, and coastal surveillance.

Can UAVs program themselves?

Not exactly, but Stanford University created a UAV helicopter that can learn to fly complex stunts by "watching" another aircraft perform those maneuvers. The helicopter gets its flight plan by recording everything that a human operator does while flying another aircraft. Then the helicopter's GPS system and avionics communicate with a ground-based computer to work out a flight plan.

How do the international laws of war impact UAV operations?

Laws of war such as the Geneva Conventions place a burden of combatants to limit collateral damage through proper target identification. In order to be able to assign responsibility for the actions of a UAV a human has to be "in the loop" - completely autonomous UAVs are not an option. In addition the person "in the loop" should be a member of the military who understands and accepts his or her role as a combatant.

Military Use of UAVs

When did the U.S. first consider using unmanned aircraft in combat?

The idea has been around for a long time. Consider the following quote from General Henry Arnold (who commanded the Army Air Forces during WWII):
"We have just won a war with a lot of heroes flying around in planes. The next war may be fought by airplanes with no men in them at all."
General Arnold may have been a little premature but that prediction is coming closer to reality all the time.

Where did the U.S. first use UAVs in combat?

During the Vietnam war RPVs (remotely piloted vehicles) equipped with video cameras flew reconnaissance missions over North Vietnam to spot troop concentrations, supply routes, and anti-aircraft sites. RPVs were also used to do electronic jamming and to drop propaganda leaflets. The RPV was launched from a C-130 and piloted by two officers aboard the C-130. Once the mission was complete the drone was maneuvered close to a CH-53 helicopter, the RPV's parachutes were deployed, and the CH-53 snagged one of the parachutes and carried the drone back to its base.

Who first used UAVs to attack battlefield targets?

During the Yom Kippur War in 1973 Syrian missile batteries in Lebanon damaged or destroyed so many Israeli fighter jets that Israel turned to unmanned aircraft research and led to their development of the first modern UAV. The effectiveness of Israeli UAVs allowed them to attack and destroy virtually all Syrian surface-to-air missile sites in southern Lebanon at the start of the 1982 Lebanon War. The success of the Israeli UAVs caught the attention of other nations and spurred the development of unmanned aerial vehicles by the military in other countries.

How fast is the use of military UAVs increasing?
U.S. Military use of UAVs has increased so rapidly over the last few years that the military is having a hard time keeping up. Flight hours over Afghanistan and Iraq more than tripled between 2006 and 2009, but only about a third of the requests by ground commanders for UAV assistance are being filled due to a shortage of aircraft and controllers.

What UAVs do the Air Force and Army fly?
The Army primarily flies RQ-7 Shadow and RQ-11B Raven UAVs and the MQ-1C Gray Eagle (an updated version of the MQ-1 Predator). The Air Force uses both the smaller UAVs like the Raven and ScanEagle and the larger UAVs like the MQ-1 Predator, MQ-9 Reaper and the RQ-4 Global Hawk.

What is the "Tier system"?
The U.S. Military uses a "Tier" system to categorize the different UAVs each service uses. For example the Air Force tier system breaks down this way:
- Tier N/A – small/micro UAVs (currently handled by BATMAV UAV)

- Tier I – Low altitude, long endurance flights (currently handled by the Gnat 750)
- Tier II – Medium altitude, long endurance (currently consists of MQ-1 Predator and MQ-9 Reaper)
- Tier II+ - High altitude, long endurance (RQ-4 Global Hawk)
- Tier III – High altitude, long endurance low-observable (RQ-170 Sentinel)

In comparison, the Army's Tier system includes the RQ-11 Raven (Tier I), RQ-7 Shadow (Tier II), and the RQ-5A/MQ-5A/B Hunter (Tier III).

How much did the U.S. use UAVs in Iraq?
During the war in Iraq, UAV missions accounted for 80% of all flights over Iraq.

How were UAVs used during Operation Iraqi Freedom?
UAVs were used extensively right from the beginning:
- Prior to the start of Iraqi Freedom Hunter UAVs were flown near Iraq's border with Kuwait to gather reconnaissance on Iraqi movements, AA sites and other information.
- Hunter UAVs were also used to locate where enemy forces were storing ammunition and weapons.
- Army Shadow UAVs were used to perform perimeter guard duty around U.S. Camps.
- UAVs helped keep the Republican Guards from making it to Baghdad to reinforce the troops there by pin-pointing the location of the

Guards units and by guiding attacks against them.

- UAVs also provided protection for U.S. convoys and ground troops, warning them of enemy activity in the area and possible ambushes or booby traps.
- Real-time UAV day and night surveillance allowed Coalition forces to detect enemy activity and mount an attack within minutes.
- Predator and Reaper missile attacks and laser targeting helped to disrupt and destroy Iraqi communications and control.

Are UAVs taking over from manned aircraft?
Yes, to some extent. In 2009 the U.S. Air Force trained more UAV pilots than fighter and bomber pilots combined. And in 2010 the U.S. bought more unmanned aircraft than manned aircraft.

Does the Air Force Academy have a UAV training program?
In 2009 the Air Force Academy started a UAV training program. Cadets flew Viking 300 UAVs over Fort Carson to get hands-on experience flying a UAV.

How many UAVs does the U.S. Military have?
As of 2009 the U.S. Military had over 7,000 UAVs, ranging from hand-launched Ravens to jet-powered Global Hawks. And that's just the beginning - within the next 20 years the Air Force expects to have entire squadrons of unmanned fighters, bombers and tankers, along with thousands of bug-sized UAVs.

What types of missions do UAVs fly in Afghanistan?

UAVs like the Global Hawk and Raven are used for surveillance and reconnaissance. Others like the Predator and Reaper carry out actual combat missions, either firing missiles themselves or illuminating targets with laser pointers for helicopter pilots.

What are the main advantages of using UAVs?

Aside from the obvious advantage of not putting aircrews at risk, UAVs are able to stay airborne for 24 hours or more carrying out surveillance over a given area without being detected. That allows analysts to study possible enemy activity over a period of time, make sure of their targets and coordinate an attack to take out those targets while avoiding civilian casualties. And even in cases where decisions have to be made in a hurry there are a number of people in the loop which should result in better decisions and fewer mistakes. There is also a major benefit for troops on the ground who get to see images relayed from UAVs overhead, showing them what's waiting around the next bend in the road or past the next building. And in situations where troops run into trouble armed UAVS orbiting the area may be able to move in quickly to take out snipers, mortar teams and other threats.

What are the main disadvantages of using UAVs?

Military UAVs currently in use are very susceptible both to advanced enemy anti-aircraft weapons and to interruptions in their communications linkage (either from natural causes or from enemy action).

As newer UAVs come into service they will be able to fly higher and faster and may include electronic counter-measures, which should make them less vulnerable to air defense systems. A possible solution to the communications problem is a manned aircraft that would serve as an airborne controller, handling one or more UAVs remotely and removing the need for long range communications links. For long-range communications, enhanced electronic security measures will help to prevent an enemy from hacking into the satellite signal. An additional problem with current UAVs is the slight time lag in communication between the UAV and its operators – a factor that should also improve as the technology involved improves. Finally, the lack of physical cues and limited peripheral vision can make a UAV more difficult to fly than a manned aircraft - a combination of experience, improved technology and training may eventually help to minimize that problem.

How does the weather affect UAVs?
Wind can definitely be a factor with small UAVs. Larger versions like the Predator are also temperamental about the weather – rain for example can soak into the carbon-fiber skin of a Predator, adding weight to an airframe that is already on a carefully balanced weight to fuel ratio. Certain weather conditions can also degrade the communications link between a UAV and its controllers.

What happens when the communications link to a UAV is broken?
When a "lost link" occurs, UAVs like the Predator and Reaper simply start to circle on a preset course and wait for the link to be re-established.

Which military UAVs carry weapons?
So far the only armed UAVs are the MQ-1 Predator and the MQ-9 Reaper.

Are Air Force UAV fliers "pilots" or "operators"?
The officers who fly armed UAVs are officially referred to as "pilots" - however some pilots who fly manned aircraft believe the title should be "operators" since UAV pilots are not physically in the aircraft. As the Air Force transitions more and more to UAVs though, it's becoming a moot point – the people who fly the UAVs do the same job and are in fact "pilots" just as if they were sitting in a cockpit.

Do UAV pilots receive flight pay?
In late 2009 the Air Force got the go-ahead to begin giving UAV pilots flight pay. It's not clear yet whether sensor operators will be able to draw flight pay, but both pilots and sensor operators will be able to wear wings for their uniforms, similar to those worn by manned pilots and aircrew members. The wings, flight pay and other incentives are part of the Air Force's shift toward a largely unmanned fleet by 2047.

Is "UAV pilot" a specific Air Force occupational code?
UAV pilots now receive an "18X" Air Force Specialty Code – it is now a "rated" career field and comes

with a six-year active duty service commitment. Sensor operators will receive the "1U" Specialty Code, requiring a 3-year active duty commitment.

Is there resistance within the U.S. Air Force to the use of UAVs?

In a word, Yes. There seems to have been a fair amount of resistance, especially at first, from the upper levels of the Air Force. However there was also resistance to the use of cruise missiles rather than strike aircraft during the Cold War and just as cruise missiles became a standard part of the Air Force, UAVs are quickly becoming an indispensible part of Air Force operations.

What is an MQ-1 Predator?

The MQ-1 Predator (made by General Atomics) is a "Tier II" medium altitude, long endurance UAV used primarily by the U.S. Air Force and the CIA. The Predator carries cameras and other sensors for reconnaissance missions and has also been modified to carry and fire two AGM-114 Hellfire anti-tank missiles or other muniitions. Powered by a gasoline engine and propeller driven the MQ-1 has a range of up to 400 nautical miles and a loiter time over target of about 14 hours. Predators have also seen civilian use for border patrols and for scientific studies.

How long can a Predator UAV stay airborne compared to an F-16?

An F-16 can stay over a target area for about an hour before it has to land or refuel from an airborne tanker. In contrast a Predator can linger over a target area for up to 24 hours and use about a tenth the amount of fuel consumed by the F-16.

22

How big is a Predator?
A Predator is twenty-seven feet long with a 49 foot wingspan and weighs about 1200 pounds without fuel or missiles.

What does a Predator UAS consist of?
A Predator Unmanned Aircraft System or UAS consists of four Predators, a ground control station (GCS) that houses the pilots and sensor operators, a satellite link communication suite, and operational personnel.

What is a "MAC"?
Most Predator and Reaper missions in Iraq or Afghanistan are actually flown by personnel in the U.S. working in a van known as a Multi-Aircraft Control system or MAC. Two officer pilots and four enlisted sensor operators in the MAC can fly up to four missions at a time. The sensor operators consoles are almost identical to the pilots', allowing the enlisted men to take control of the UAV. The sensor operaters fly the UAV inside an airspace defined by the pilot and generally the pilots only take over during an emergency or to fire weapons.

How many people are in a Predator squadron?
A typical Predator squadron consists of 350 to 400 people, including 27 pilots.

What exactly is an MQ-9 Reaper?
The Reaper, also referred to as the Predator B, is a medium-to-high altitude UAV whose primary objective is to act as a hunter/killer against targets of opportunity. Its secondary mission is to act as a

surveillance and reconnaissance platform, using various sensors to relay real-time data to field commanders and intelligence specialists. The Reaper has a complete sensor suite including an infrared sensor, a color/monochrome daylight TV and an image-intensified TV. Reapers can carry a variety of weapons including the AGM-114 Hellfire missiles, GBU-12 Paveway laser-guided bombs, the AIM-9 Sidewinder air-to-air missile, and the GBU-38 JDAM (Joint Direct Attack Munition). In addition the MQ-9 has a laser rangefinder/designator giving it the capability to designate targets for laser-guided munitions.

How big is a Reaper?
An MQ-9 Reaper is thirty-six feet long with a 66 foot wingspan and weighs about 4900 pounds without fuel, missiles or bombs. The Reaper also flies at about 300 miles per hour (compared to about 100 miles per hour for the Predator).

What does a Reaper UAS consist of?
A typical Reaper UAS consists of multiple UAVs, a ground control station, communications equipment, spare parts, operations and maintenance personnel. Like the Predator, the Reaper is crewed by a pilot and a sensor operator.

How do Predators and Reapers compare to fighters like the F-15 or F-16 costwise?
The MQ-1 Predator costs about 7 ½ million dollars and the Reaper runs about 13 ½ million, compared to 70 or 80 million for an F-16.

Are all Predators and Reapers used in Afghanistan and Iraq flown by operators in that country?
Predator and Reaper missions are actually flown by operators stationed in the U.S. at places like Creech Air Force Base near Las Vegas. However there is a lag of about 1 ½ seconds in the satellite signal from overseas which is too long a delay for a pilot trying to take off or land, so UAVs are still launched and landed by personnel in Iraq or Afghanistan.

How many Predator and Reaper missions are flown every day in Iraq and Afghanistan?
As of 2009, an average of about 30 unmanned combat air patrols were being flown every day.

What kind of sensors does a Predator or Reaper have?
Predator and Reapers contain a full suite of sensors: electro-optical, infrared, video, and laser/target markers.

What type of armament does a Reaper carry?
A weapons loadout for a Reaper normally consists of four Hellfire missiles and two 500 pound GBU-12 laser-guided bombs.

How long can a Predator stay airborne?
Predators often stay in the air for up to 24 hours (with a shift change in operators).

How many Predators have crashed to date?
As of the fall of 2009, approximately one third of the 200 Predators delivered to the U.S. Air Force had crashed, either because of aircraft malfunction or

operator error. However, better documentation and more intense training programs now in operation appear to be helping to reduce that percentage.

What is a Global Hawk?

The RQ-4 Global Hawk is a high-altitude, long endurance surveillance and reconnaissance UAV which can provide area coverage in all weather, day or night. The Global Hawk UAS consists of the aircraft, its sensors, a ground station, data link, and trained personnel. Different versions of the RQ-4 are being developed and fielded in "capability blocks". Block 10 is currently operational and carries imagery intelligence (IMINT) sensors.

Note: Block 20 will include a larger aircraft with increased payload capacity and improved IMINT. Block 30 will have both the improved IMINT sensors and integrated signals intelligence (SIGINT) sensors to achieve simultaneous, multi-INT coverage. Block 40 will comprise a separate set of aircraft employing an advanced radar sensor to provide improved IMINT and surface/air moving target information.

How much does a Predator, Reaper or Global Hawk cost?

Military UAVs like the Predator, Reaper and Global Hawk generally cost tens of millions of dollars. According to one report a Predator costs about 4 to 5 million dollars, a Reaper about $11 million (half as much as an F-16) and a first-generation Global Hawk runs about $35 million. As of a couple of years ago, the cost of a second-generation RQ4-B Global Hawk had increased to around 56 million dollars. NOTE: cost estimates for military UAVs can vary widely depending on whether the figures are for the UAV only or for an entire UAS, including ground control and communications equipment.

Does the U.S. Army fly Predators?
As of 2009 the Army has units using the MQ-1C Gray Eagle, an enhanced version of the MQ-1B Predator.

What other UAVs does the Army use?
Two of the primary UAVs the Army uses are the RQ-11B Raven and the RQ-7 Shadow.

What is an RQ-7 Shadow?
The RQ-7 Shadow UAV is used by the U.S. Army and Marine Corps to carry out reconnaissance missions. It's launched from a trailer-mounted pneumatic catapult and recovered using arresting gear similar to that on an aircraft carrier. The RQ-7 is a little over 11 feet in length with a wingspan of 14 feet and weighs about 185 pounds when it's empty. It has a maximum speed of 135 miles per hour, a range of 68 miles and a maximum altitude of 15,000 feet.

What does an RQ-7 Shadow UAS consist of?
A Shadow UAS includes four UAVs, two fixed ground control stations (GCSes), one portable GCS, one launcher, two Ground Data Terminals (GDTs), one portable GDT and one Remote Video Terminal.

How many people are needed to operate a Shadow UAS?
Currently it takes 22 people to fully man a Shadow UAV system.

Are other countries using UAVs?

More than 40 countries currently have UAVs in their aircraft inventory and more are on the way. For example, China has at least 25 models of UAVs currently in service including the Combat Eagle stealth UCAV and the Thunderbolt UCAV. One day in the not-too-distant future, U.S. UAVs may be engaging enemy drones in unmanned aerial dogfights.

Could UAVs be used in terrorist attacks in the United States?

At some point UAVs may become a major concern in regards to terrorist attacks. For example small or micro UAVs could be used to track people and provide information to allow terrorists to set up an ambush – or a small UAV filled with a deadly toxin could be flown into a statium or other crowded area and the toxin released by remote control.

How accurate are UAV attacks?

There have been questions raised about the accuracy and targeting of UAV attacks. Israeli UAVs armed with missiles killed 48 Palestinian civilians in the Gaza Strip in 2009; those casualties were blamed on the failure of operators to distinguish civilians from combatants. Another problem with drone strikes is the weapons involved – for example the MQ-1 Predator is normally armed with 100 lb Hellfire missiles which were designed to take out tanks and to attack hardened bunkers. Smaller weapons such as the Raytheon Griffin and the U.S. Navy's Spike missile are being developed to offer a weapon load capable of a more surgical attack. Along with increased accuracy, missiles like the Griffin also allow the Predator A to carry six missiles rather than two of the much heavier Hellfires.

What kind of engine do most UAVs use?
Smaller UAVs are generally powered by small rotary or two-stroke piston engines while some of the large UAVs like the Reaper are jet-powered.

Do military UAVs crash more often than manned aircraft?
Actually, despite the fact that UAVs fly more hours than manned fighters or bombers, their safety record is about the same. UAVs don't have redundant systems like manned aircraft (since crew safety isn't involved) and that contributes to the number of UAV crashes, as do pilot error and control signal interruptions. In addition because of the demand for UAVs the software itself has some weaknesses – for example, one Predator was lost because the pilot inadvertantly managed to erase the onboard RAM (Random Access Memory) making the UAV impossible to control.

Do military UAVs fly according to a program or are they actually piloted by a human?
It depends on the particular UAV. Some, like the Global Hawk, carry out their missions according to programmed instructions – ground controllers can transmit updated programming to the UAV but they can't pilot it directly. Other systems fly on autopilot most of the time (except for takeoff and landing), but can be controlled directly by a human pilot when necessary.

What are ground control stations like?

Military control stations are normally mobile and can be in a building, a vehicle, or another aircraft. Most of them consist of two workstations with one used by the pilot and the other by a sensor specialist (although either workstation can be used to pilot the UAV in case the pilot's console has a problem). In some cases there is also a third station for a "mission intelligence specialist" whose main job is to coordinate with friendly ground forces. The station's electronics package normally allows the UAV to be controlled either by direct radio communication or by satellite link when the UAV is beyond range of the line-of-sight data link.

What does a Predator control station look like?

The pilot and sensor operator sit in front of an array of monitors. There are two primary monitors – one shows the video feed from the Predator's cameras overlaid with a display of the horizon, the Predator's altitude and other vital information while the other monitor displays a graphic image of the UAV overlaid on a satellite map of the area.

How does a UAV operator make a Predator or Reaper change directions?

The "pilot" draws a line on the screen of his monitor and the UAV alters its pre-programmed flight path, following the changes drawn on the monitor.

Do you have to be an experienced pilot to be allowed to pilot UAVs?

The Army and Marines rely primarily on enlisted personnel who have never flown, but they are only dealing with small, unarmed UAVs. Until recently the

Air Force only allowed pilots who had flown real aircraft (fighters, helicopters, heavy transports, etc.) to pilot the larger UAVs, particularly the Predator and Reaper armed UAVs. However, due to the pressure to add more pilots the Air Force began a new program in 2009 to train officers with no prior time in a cockpit as UAV pilots. The program involves six months of screening and basic flight training followed by a nine-week initial qualification course including forty hours in a simulator and at least nine actual flights. So far results have indicated that non-pilots do about as well as experienced fliers.

How receptive are Air Force pilots to being assigned to fly UAVs?
There seems to be a mixed reaction, with pilots who have been flying helicopters or "heavies" (such as transports or tankers) being more receptive than fighter pilots to transitioning to UAVs.

Author's Note: Years ago when I was in the Army I spent about 18 months as liason to an Air Force squadron flying Army requested missions. The aircrews I worked with flew Phantoms and were bright and dedicated officers. However, my feeling is that neither the pilots or backseaters would have been happy being taken out of the cockpit and put in a UAV control station. I suspect that's still true of fighter pilots today – they have a dangerous job (even if they're not flying in combat) and I think part of what attracts them is testing themselves physically and mentally every time they go up – that and the "need for speed".

What are the basic duties of an Army UAV operator?

Army personnel assigned to the 15W MOS (Military Occupational Specialty), Unmanned Aerial Vehicle Operator perform the following duties:

- Plans, prepares and helps carry out air reconnaissance missions.
- Operates mission sensors to locate and identify targets.
- Deploys and redeploys the Tactical Unmanned Aerial Vehicle (TUAV) ground and air system.
- Operates and performs operator level maintenance on communications equipment, power sources, light and heavy wheeled vehicle and some crane operations.
- Launches and recovers the UAV.
- Performs pre-flight, in-flight, and post-flight checks and procedures.
- Directs emplacement of the ground control station and launch and recovery systems.
- Supervises and assists in air frame repair. Coordinates replacement of parts and other necessary supplies.

What kind of training is required to fly an Army Raven UAV?

Army troops have to complete an 80 hour course to qualify as a Raven operator. Setting up a mission properly is important, but according to the trainees the hardest part of the course on the hand-launched Raven is ... mastering the technique of hand launching.

Where is the Army's UAV training center?

Fort Huachuca in Arizona is home to the world's largest UAV training center. Over 350 employees of General Dynamics teach at "Black Tower" training students on the Hunter, Shadow, Gray Eagle and Warrior Alpha systems. Students at the Black Tower facility spend at least half their time flying simulated UAVs – in some cases soldiers complete 100 hours of simulation compared to just 10 hours of flying an actual UAV. Fort Huachuca is not only the largest UAV training center in the world it's also the only place in the world where unmanned aircraft have priority over manned aircraft.

Where is the Air Force's UAV training center?

Currently the main training center for the U.S. Air Force is at Creech Air Force Base in Nevada (Note: until 2005 Creech AFB was an auxiliary airfield for Nellis AFB). Pilots at Creech Air Base also fly most of the Predator and Reaper missions in Iraq and Afghanistan (although launches and landings – which require quick response - are handled on site by UAV operators in Iraq and Afganistan, due to the slight lag time in communications between there and the U.S.).

Do Air Force UAV pilots have to meet the same standards as manned aircraft pilots?

At present the Air Force may relax some physiological requirements, but UAV pilots must meet the same requirements for eyesight as manned aircraft pilots and the same academic standards, including passing the Test of Basic Aviation Skills and the Air Force Officer Qualification Test.

What is Air Force UAV operator training like?

First of all, except for some small short-range reconnaissance drones, UAV pilots have to be qualified officer pilots. Sensor operators however are normally enlisted personnel and have the ability to take control of the UAV when necessary. Flying a UAV requires the same basic skills as flying any aircraft. However the emphasis for UAV pilots and sensor operators may be more on information processing than on actual flying. Semi-autonomous UAVs can handle a good deal of the mechanical part of flying on their own – the primary job of the operators is to assimilate the information being passed to them and make appropriate decisions based on that data.

A good deal of the training involves learning how to use the UAV's sensors to best advantage. For example, during mid-day in warm weather, a Predator's infrared imaging can tell you if a vacant parking lot has had vehicles parked there recently. The spaces where the vehicles were parked will be cooler than the rest of the lot since they were shaded from the direct sun.

What characteristics make a good military UAV operator?

- Proficiency at flying the UAV
- The ability to concentrate for considerable periods of time when very little is happening – to stay alert and watch carefully for any items of significance.
- Talent at developing tactics "on the fly" - the ability not only to pick out obvious targets, but

to do "pattern analysis", to study the enemy's habits, where they store supplies, what routes they like to take, where they generally come from and where they go.

- The ability to work closely with others – especially since UAV controllers usually work as a two-man team, with one person doing the piloting and the other analyzing the imagery from the UAV.

What is the hardest part of piloting an armed UAV?

The most difficult thing about flying a Predator or Reaper is the same as with a jet fighter – deploying the weapons. Trying to coordinate with troops on the ground who want you to attack a target that has friendly troops and/or civilians within 50 feet from it can be extremely challenging.

What advantage does a UAV pilot have over a fighter pilot?

Aside from personal safety, a UAV pilot may spend up to three years flying missions over a certain area compared to a few months for a fighter pilot. That allows the UAV pilots to become much more familiar with the territory they're observing – which makes it easier for them to spot changes in behavior, increased traffic in a certain area, or other things that may indicate terrorist activity. Familiarity with the people and daily life in the operational area can also help differentiate between real insurgents and innocent civilians.

What are some disadvantages of being a UAV pilot or sensor operator?

One disadvantage is the amount of information UAV operators have to deal with. Not only are there readouts from multiple sensors, there are the flight controls, the communications link, and the need to coordinate with ground troops or other personnel. Also, with the current shortage of UAV pilots, workload presents an additional problem. Constantly changing schedules in particular can upset a pilot or sensor operator's internal clock, causing mental stress and fatigue. An additional concern for UAV crews is the feeling of impotence in cases where they can see events happening that are putting troops or civilians in danger and they aren't in a position to do anything about it. There is also some psychological stress in being involved in battlefield operations for hours and then going home to normal family life – that constant adjustment and the lack of fellow soldiers to talk with during off-duty hours can be difficult.

What is the "multiple-mission model"?

Some experts feel that the solution to the shortage of UAV pilots is to advance the technology that allows a single pilot to control multiple missions. The multiple-mission model is a framework that may lead to a pilot being able to conduct a dozen or more missions simultaneously.

How similar is flying a UAV mission to flying a jet fighter?

Very similar according to pilots who have flown both types of aircraft. For example every Predator mission starts with mission planning, including weather, route and intelligence briefings, the same as for a manned flight. Once the mission planning is complete the UAV crew moves to the ground control station and goes through a pre-flight checklist just as they would for a manned fighter, except that the checklist is done over the radio with the UAV ground crew. The control station itself is arranged like a cockpit, with the pilot sitting in the left-hand seat and a sensor officer in the right-hand seat. The ground station controls include a stick and rudder, but with multiple computer screens taking the place of most of the other buttons, switches and warning lights that you would find in a normal fighter cockpit. Piloting the UAV also is also very similar to flying a fighter according to the air crew, with the main exception being that there aren't the same physical cues as in a manned aircraft – no sound or vibration and limited peripheral vision (a fact which can actually make the UAV somewhat harder to fly and land). Also, UAV support and maintenance procedures are about the same as for a jet fighter and require a ground crew of about 50 people who are responsible for the weapons, electronics, radio and video gear, airframe and engine.

What is involved in mission planning?
Mission planning is the creation of a flight plan that takes into account:
- Terrain
- Weather
- Aircraft performance capability and configuration
- Weapon and/or cargo payload
- Fuel requirements

- Assessment of the route based on any known enemy threats
- Checking of the flight route and timing versus other aircraft in the area to avoid conflicts

What is the AFMSS?
The Air Force Mission Support System provides automated mission-planning support. As far as UAVs are concerned the system has been used to flight plan Global Hawk missions.

What are the advantages to a flight crew of flying a UAV rather than a manned aircraft?
The primary advantage of a UAV is the lack of outside distractions – not only the operators' knowledge that they're not personally in danger but also the lack of gut-wrenching high-g maneuvers or other physical stress on the crew. Also, many UAVs can stay airborne for long periods of time and in those situations a new crew can take over after a certain number of hours, ensuring that crew members stay alert and ready to respond quickly to any situation.

How much does it cost to train a fighter pilot versus a UAV pilot?
According to an article in the Air Force Times it takes more than a year and over 2 ½ million dollars to train a fighter pilot, compared to 20 weeks and $135,000 to train a UAV pilot.

Can a military UAV maneuver like a jet fighter?
It depends somewhat on the type of UAV. The MQ-1 Predator is powered by a small piston engine (115

hp) and attempting hard turns at high speed can send it out of control. Later models like the MQ-9 Reaper have a much more powerful turbofan engine (about 950 hp) and are capable of more high-speed maneuvers. However no current UAV can match the ability of a fighter like the F-16 or F-22 to accelerate, turn, climb, or dive.

What exactly happens when a UAV fires a missile?

When the pilot fires a missile the sensor operator uses a laser designator to guide the missile to its target. If the target is moving the sensor operator has to keep it in the crosshairs of the laser designator until the missile gets there, which can be 30 seconds or more.

Who gives permission for a UAV to fire on a target?

If the UAV is flying a combat surveillance mission the personnel in the ground control station normally make the decision to fire. If the UAV is performing a CAS (Close Air Support) mission in support of troops engaged with the enemy, then the JTAC (Joint Terminal Attack Controller) on the ground in Iraq, Afghanistan or wherever has to give clearance to fire.

Are there UAVs that can take off and land on an aircraft carrier?

There are smaller UAVs that can be employed on a carrier, but in February 2011 the U.S. Navy conducted the first flight of the X-47B, an F-18 sized UAV that will be carrier-capable. The X-47B is designed for stealth and is meant to serve as a carrier-launched robotic strike aircraft. Since the X-47B is jet powered it can fly much faster than prop-driven UAVs like the Predator and Reaper and should be able to carry out a range of missions. Currently the X-47B is slated for carrier trials sometime in 2013.

What is the UXV Combatant?
BAE Systems is working on designing what amounts to a UAV carrier for the British Royal Navy. The warship is expected to displace approximately 8,000 tons and feature two decks for launching UAVs, VTOL (Vertical Take Off and Landing) aircraft and helicopters. The UXV Combatant reportedly will also be able to launch unmanned submarines and medium-range missiles and is expected to enter service with the Royal Navy sometime after 2020.

Can an enemy gain information from our UAVs?
There have been several instances where insurgents have gained intelligence of their own from our UAVs. In one case, when one U.S. combat division would end its tour of duty, it would ground all its UAVs until the new division moved in and took over. The insurgents figured out that pattern and took advantage of the "down time" to carry out operations while they knew they weren't being watched.

Are there any all-UAV fighter squadrons?
Yes, the New York Air National Guard 174th Fighter Wing began transitioning from F-16 fighters to MQ-9 Reapers in 2008 and became the first all-UAV attack squadron.

Why were baby wipes important in Iraq?
Baby wipes were used to clean the sand and other grit and grime off of the Army Hunter and Shadow UAVs in order to keep them flying.

Has an enemy unit ever surrendered to a UAV?
Oddly enough, yes. During the Gulf War Iraqi Army forces surrendered to the UAVs of the USS Wisconsin.

Has a UAV ever been in a dogfight with an enemy fighter?
A UAV has engaged a jet fighter on at least one occasion. In 2002 a U.S. RQ-1 Predator exchanged missile fire with an Iraqi MIG-25, with the MIG's missile destroying the Predator.

Are there UAV test pilots?
The U.S. Military just recently appointed its first drone test pilot, Captain Nicolas Helms, stationed at the USAF test pilot school at Edwards Air Force Base in California. His job will be to "push the envelope" and establish the performance limits of various UAVs.

What is the Air Force's "UAS flight plan"?
In 2009 the U.S. Air Force approved the "Air Force Unmanned Aircraft Systems Flight Plan (2009-2047)", outlining a strategy for integrating UAS operations with all Air Force core functions and joint force priorities.

What is an ARSS?
The Autonomous Rotorcraft Sniper System or ARSS is one of the U.S. Army's answers to the problem of

fighting guerrillas who can attack and then hide among the local citizenry. The ARSS is a small unmanned helicopter (a Vigilante 502 UAV) equipped with a .338 sniper rifle. An autopilot handles the business of flying the UAV while a human operator on a remote monitor fires the rifle. To keep the cost down a good deal of the hardware in the ARSS is off-the-shelf, including an Xbox 360 video game controller that operators use to control the turret. Other weapons are also being considered for use with the ARSS including the M-249, an AA-12 shotgun, or non-lethal weapons.

What are "Micro UAVs"?
Miniature UAVs are an area getting increasing attention. Micro vehicles range from UAVs that can be carried by an individual infantryman to man-portable models that can be carried along with an infantry unit and launched like an anti-aircraft missile.

What is the "LMAMS"?
The LMAMS (Lethal Miniature Aerial Munition System) is a man carried missile with a number of UAV characteristics, including loitering time and a hand-held ground station.

I've read that the last manned fighter pilots have already been born ... is that true?
Never say never ... but very unlikely. There will always be a need for humans in the cockpit. When line-of-sight and satellite communications fail, you have to have the option of having human "eyes in the sky".

What will the next generation of military UAVs be like?

New types of UAVs are being introduced almost every week, so it's hard to define a "next generation". However there are some interesting models currently under development:

- The U.S. Navy is working on a bat-winged bomber (designated the X-47B) designed to take off from an aircraft carrier. The X-47B is expected to be part of the new generation of UAVs that can evade radar and fly much faster than propeller-driven Predators and Reapers. The X-47B is also an example of a shift in emphasis from the production of short-range tactical strike UAVs only to the development of longer-range strategic bombers.

- Aurora demonstrated their "Skate" UAV at the Farnborough International Airshow in July 2010. The Skate is a little larger than a pizza box and is made from a Styrofoam-like material. It has two electrically powered tiltrotor propellers, onboard sensors, and a downlink capability that allows it to connect to a Windows or Linux PC. The tiltrotors let the Skate switch quickly between vertical and horizontal flight, somewhat like the V-22 Osprey that the U.S. Marines use.

- AeroVironment's Global Observer high altitude, long endurance (HALE) UAV made it's first test flight under hydrogen power in January 2011. Global Observer has a 175 foot wingspan and can use solar power and hydrogen cells to drive four electric motors, allowing it to stay aloft for up to a week at altitudes of 55,000 to

65,000 feet. It's intended to provide satellite-type coverage of a given area at a fraction of the cost.

- Electric UAVs kept aloft by laser power-beaming technology may provide almost unlimited flight time.
- The Defense Advanced Research Projects Agency (DARPA) is supposed to be working on the design of a UAV that will be able to stay aloft without any maintenance for up to five years.
- Some experts even predict that in the not-too-distant future UAVs capable of vertical take-offs and landings will detect enemy activity, land and release small unmanned robots armed with various types of weaponry. The robots (possibly guided from the UAV once it's airborne again) will carry out the actual attack on the opposing forces, taking the place of human troops on the ground.

Civilian Use of UAVs

Who can fly a UAV in civilian areas in the U.S.?

The Federal Aviation Administration (FAA) classifies UAVs as civil aircraft and requires them to be operated by pilots certified under FAR Part 61 (the section of Federal Aviation Regulations that cover pilot certification and standard flight school operations).

What are some non-military jobs where UAVs are being used?

- The U.S. Customs and Border Protection agency currently uses MQ-9 Predator B UAVs to patrol the border between the United States and Mexico. Infrared cameras mounted in the Predator's nose transmit images by microwave link back to Fort Huachuca in Arizona. When an observer there spots something suspicious the information is relayed to a Border Patrol Station, which sends a Blackhawk helicopter to investigate. If necessary the Predator can then illuminate people or vehicles on the ground with a laser designator to guide the Blackhawk directly to the target.
- Scientific research that involves obtaining data from remote or hazardous locations, such as flying into the center of a hurricane to get meteorological data. For example, in 2005 the National Oceanographic & Atmospheric Administration (NOAA) flew an Aerosonde UAV into the heart of tropical storm Ophelia, gathering data in winds of 175 mph.

- Cargo carrying – helicopter or blimp UAVs can haul various types of payloads at a low cost.
- Fighting wildfires – Modified Predator Bs from NASA have been used to map forest fires and provide thermal imagery that can be passed on to firefighters on the ground. The information gained from the imagery can help locate the areas of maximum danger to firefighters and to analyze the spread of the fire.
- Monitoring oil and gas pipelines – above-ground pipelines can be carried out by digital cameras mounted on UAVs.
- Crop dusting – UAVs have been used in Japan for crop dusting.
- Celebrity watching – Supposedly a celebrity-photo agency is developing a camera-equipped paparazzi drone to track and photograph celebrities.
- Search and Rescue – The U.S. Coast Guard has tested the Bell Eagle Eye Tr-916 as a ship-based search and rescue platform.
- Mapping – Some testing has been done with an eye to using very small, light UAVs with video cameras to do urban mapping somewhat like Google Earth and Google Maps.
- Environmental monitoring – In October 2010 a Meridian UAV at the University of Kansas made its first flights over the Arctic, gathering data to monitor the movement of polar ice.
- Disaster response – UAVs are projected to become an integral part of response to natural disasters in the future – unmanned aircraft with infrared (heat) sensors could be life-

savers, spotting victims trapped in damaged buildings or homes.

Have UAVs actually been used in civilian search and rescue operations?
UAVs were used successfully in the aftermath of the 2008 hurricanes that hit Louisiana and Texas. Photos from before and after the storm highlighted areas of damage and helped to direct rescue efforts. In addition, an RQ-4 Global Hawk was sent from Beale AFB to Haiti in January 2010 to assist in the humanitarian aid operations there following a 7.0 magnitude earthquake.

Have UAVs been used for security purposes?
India used UAVs during the 2010 Commonwealth Games to provide surveillance for the Games Village and the different venues. The primary purpose for the use of UAVs was to prevent a possible airborne attack by a terrorist group using paragliders.

Won't civilian UAVs be a danger to other aircraft?
Currently the FAA restricts operation of UAVs to certain specific areas to avoid the possibility of collisions with planes or helicopters. However systems under development to use a combination of video cameras and sophisticated image processing to give UAVs the equivalent of a human pilot's "see and avoid" ability to search for incoming aircraft.
Note: The FAA's transition from radar-based air traffic control to a satellite-based system may also make UAV flights in civilian airspace less of a concern since under the satellite system (known as

ADS-B) aircraft will continuously broadcast their position.

What are the current FAA restrictions on flying a UAV in U.S. Airspace?

Civilian UAVs are limited to line-of-sight operation, altitudes of less than 400 feet and are barred from flying near airports.

What is a "COA"?

The Federal Aviation Administration has issued Certificates of Authorization (COAs) in certain cases to allow the use of UAVs in civilian airspace. The number of COAs issued has been a trickle though compared to the demand.

Does weather pose a significant problem for commercial UAVs?

Weather conditions could pose a problem for some current commercial models. Those UAVs that only have electro-optical sensors could be "blinded" by heavy cloud formations – military models with synthetic aperature radar (SAR) can see through cloud formations and weather. Onboard weather sensors and de-icing gear are also important to make sure the weather doesn't pose a threat for commercial UAVs.

What kind of education do you need to be a civilian UAV pilot?

A four year college education is helpful, but with or without a college degree you will probably need a private pilot license, a commercial pilot license and a few hundred flight hours as the pilot in command.

Is there anywhere I can see what's involved in getting a private pilot license?

There are several good books available on getting a private pilot license:

- "The Complete Private Pilot" by Robert E. Gardner
- "Your Pilot's License" by Jerry Eichenberger
- "The Pilot's Manual: Ground School: All the Aeronautical Knowledge Required to Pass the FAA Exams and Operate as a Private and Commercial Pilot" by Aviation Theory Centre Ltd.

Or you can go to:

http://www.free-online-private-pilot-ground-school.com

for a look at the basics of flying an airplane, from navigation to flight instruments to airplane aerodynamics.

How much can a pilot with UAV experience earn as a civilian UAV pilot in the U.S.?

Currently experienced UAV pilots can start at salaries as high as $100,000 a year. Salaries of $75,000 to $85,000 are possible with very limited UAV experience if you have a degree in a field like engineering or telecommunications. As more UAV pilots return to civilian life from the military, future salaries may depend on how quickly the civilian market grows (which will depend heavily on what changes are made in FAA regulations and how quickly they are made).

What is the Parrot AR.Drone?

An early entry in the civilian UAV market, the Parrot AR.Drone is a small four-bladed helicopter controlled via iPhone over a wireless network. The Parrot features two cameras and can potentially fly at altitudes up to 160 feet, beaming back video form the cameras. You can actually buy a Parrot drone online at several locations including amazon.com.

What is the "SenseFly Swinglet CAM"?

A small RC drone – basically a flying wing with stabilizers, a rear propeller, and a small on-board camera. You simply set up your GPS coordinates for the area where you want to get some aerial imagery, shake the Swinglet to start it up and then let it go.

What is the "Air-Mule UAV"?

Urban Aeronautics has developed a UAV called the "Air-Mule", designed to assist in medical evacuation. The Air-Mule will be able to carry medical equipment and fly into hard-to-reach places utilizing it's ability to take off and land vertically, similar to a helicopter.

Where can I learn to fly a UAV?

- UAV Flight School (www.uavflightschool.org) offers non-automated UAV/UAS flight training, including classroom and simulator time as well as actual flight time with a sUAS. Material covered includes procedures for aileron, engine or elevator failure, use of flaps, crosswind landing techniques, proper glide slope approach and proper use of the throttle.
- Both the University of North Dakota and Kansas State University offer four year degree

programs in UAV pilot training and Unmanned Aircraft Systems.

- Embry-Riddle Aeronautical University has also created a new minor in Unmanned Aircraft Systems at their Daytona Beach, Florida campus. Courses include Unmanned Aircraft Systems, Unmanned Aircraft Systems Operations and Cross-Country Data Entry, Operational Aspects of Unmanned Aircraft, UAS Robotics, and Unmanned Sensing Systems.
- As a first step you could try the 2nd R/C Flight School (http://www.2ndrcflightschool.com) which teaches you how to fly radio-controlled model planes or enroll at any certified flight school that can help you get your private pilot's license.

What if I want to make a UAV of my own?
Try this webpage:
http://hacknmod.com/hack/make-a-uav-spyplane-using-the-arduino/
The tutorial shows you how to build your own UAV using a handful of servos, a battery pack, a GPS unit and the Arduino programmer.

UAV Photo Gallery

UAV Ground Control Station

MQ-1 Predator in flight

MQ-9 Reaper

RQ-4/MQ-4 Global Hawk

RQ-11 Raven (hand-launched)

MQ-8 Fire Scout

USAF UAV Pilot's Wings

RC Models and UAVs for Hobbyists

Note: Learning to fly radio-controlled model airplanes is frequently mentioned as a good way to introduce the basic techniques involved in flying UAVs.

What types of engines do RC aircraft use?
RC aircraft (with the exception of gliders and other unpowered models) use electric motors, internal combustion engines, or jet engines.

What is a "glow plug"?
Glow plugs are used in model airplane engines and are similar to sparkplugs. A battery is used to heat the glow plug initially, then fuel is fed into the engine and the glow plug ignites the fuel just like the spark plugs in a automobile engine.

Why are RC internal combustion engines referred to as "nitro" engines?
The standard fuel for model airplane engines consists primarily of methanol with a certain percentage of nitromethane mixed in. That mixture is suspended in an oil base, commonly castor oil or a synthetic oil of some type. The term "nitro" comes from the volatile nitromethane part of the fuel.

What types of RC aircraft are there?
Basic types of RC aircraft include:
- Trainers – Trainers are designed to help you learn how to fly. They normally have the wing

on top for maximum stability and are powered by electric motor or by glow plug engines.

- Sport planes – Sport airplanes make up a large part of the RC airplane market. They can be any size or shape and are better at performing aerobatic maneuvers than trainers.
- Warplanes – Warbirds are a popular type of RC aircraft and generally are built for higher performance than trainers or sport models.
- Aerobatic models – These models are built specifically for performing aerobatics and are generally low or mid wing planes with powerful motors.
- Float planes – Float planes are designed to land on water, but require a good deal of skill to fly.
- Helicopters – Single and dual rotor helicopters are a very popular variety of RC aircraft and tend to be fairly easy to fly.

Vintage airplanes, jets (powered by ducted fans or miniature gas turbine engines), gliders, and blimps are also available.

How hard is it to learn to fly an RC airplane?

Most RC enthusiasts will tell you that it's harder than learning to drive a car. Flight simulators can be a good way to get started without having to worry about crashing an expensive RC model.

What are "control surfaces"?

The control surfaces on a model airplane are the parts of the plane that control its movements – the ruddeer, elevator, ailerons, etc.

What are "servos"?

The control surfaces on a model plane are actually moved by electrical impulses transmitted through the radio to small motors or "servos" in the airplane. The servos physically move the rudder, elevator or other control in reponse to those commands.

What is the most important step in getting a model airplane ready to fly?

Probably the most important thing to do is to adjust the "trim settings" for the airplane. Models that aren't trimmed properly can be almost impossible to fly. RC transmitters normally come with trim controls to allow you to adjust the airplane's servos and control surfaces so that the plane will fly straight and level with the radio controls in the centered or neutral position and your hands off the radio.

What is "FPV flying"?

FPV stands for "First Person View" and refers to flying a model aircraft as if you were on board through the use of a miniature video camera and RC relays in the aircraft.

What is an "ARF" RC model airplane?

ARF stands for "Almost Ready to Fly" - ARF models generally require just a few additional items to be flight ready.

What rules apply to RC aircraft in the U.S.?

Federal Aviation Administration (FAA) Advisory Circular 91-57 states that radio-controlled model aircraft must stay below 400 feet altitude and that if you're flying your model with 3 miles of an airport you need to notify the airport. Considering the small size, weight and limited capabilities of most RC models, current FAA regulations are fairly relaxed. However, that may change due to the rise of RC models that are basically UAVs.

Where can I find RC UAV kits?

- Http://www.nitroplanes.com/4eluavrqprra.html - 4-channel electric UAV RQ/MQ-1 Predator 63 inch radio remote controlled RC spy plane ARF ($62).
- http://www.rchelicopterselect.com/uav-rq-1-predator-4-channel-electric-rc-airplane-kit.html – RQ/MQ-1 Predator 4-channel electric RC airplane kit ($92.99).
- http://nitroplanes.com/mq9.html – PROJET MQ-9 Reaper 98 inch 4-channel brushless remote control RC UAV ARF kit V2 ($139.95). Additional options include a choice of FPV camera, charger and flight simulator.

What rule changes may be in store for RC aircraft?

According to recent (Feb. 2011) reports, the FAA may be about to impose much stricter regulations regarding the flying of RC model aircraft. Spurred by the growing number of sUAS (small Unmanned Aircraft Systems), new regulations may include a 55 pound weight limit, a ban on gas turbine engines, and an expanding of the "no fly without permission" rule on flying near airports, increasing the range from a 3 mile radius to 5 miles or more.

What about RC flight simulators?

Pilots generally start out by taking flying lessons; if you can find an experienced RC pilot to teach you that may be the way to go, but failing that, RC flight simulators can help you "earn your wings". A good quality RC simulator will consist of some very powerful software and may also include a realistic transmitter interface controller. Popular flight simulators like RealFlight and Phoenix feature a variety of aircraft to choose from, including helicopters, prop planes, fighter jets, airliners and gliders. Some simulator software even includes a virtual flight school where real-world pilots teach you the basics, from take-off and landing to advanced aerobatic maneuvers.

Is there a good RC flight simulator for $100 or less?

There are several. One of the best is RealFlight Basic which sells for right around $100. It comes with dozens of aircraft models to choose from, photo-realistic airfields, and a transmitter that has all the proper controls and functions and plugs into a USB port on your computer.

What are some other RC simulators?

- Manuel's Radio Control Flight Simulator – A free simulator that focuses on the basic principles of flight control rather than fancy graphics
- ClearView RC Flight Simulator – A commercial simulator with a demo version (limited to two airplanes and a helicopter) that lets you try out the software. Buying the full version adds a

host of other aircraft and even more user-submitted models can be downloaded from the website. You also have the option to fly in multi-player mode over the internet.

- RC Plane Master – This flight sim focuses on airplanes. The package includes a controller that plugs into a USB port and 12 aircraft, including electric, glow-plug and jet-powered models.
- E-Sky 4 Channel Flight Simulator Training Kit – The kit contains the software on a CD along with an E-Sky transmitter that plugs into your PC's USB port. E-Sky Flight Simulator is a low-cost way to get an introduction to RC flying.

What are some websites for RC aircraft?
- Http://www.rc-airplane-world.com
- http://rcvehicles.about.com
- www.nitroplanes.com
- www.rc-airplanes.com
- www.rcplanetalk.com (forum)
- www.hobby-lobby.com/radio_control_airplanes_196_ctg.htm

When does a radio-controlled model become a UAV?
Technically, a model aircraft becomes a UAV when it goes out of sight of the operator.

Are there any online UAV simulators?
http://www.airforce.com/interactive/uav/index.html offers a very basic simulation that provides a rough idea of what it's like to fly a UAV.

UAV Organizations

Are there any national or international UAV organizations?

Here are two:

- Association for Unmanned Vehicle Systems International (www.auvsi.org) – AUVSI is the world's largest non-profit organization (over 6,000 members in 55 countries) promoting the use of UAVs. Issues a monthly magazine "Unmanned Systems" to members.
- Unmanned Aerial Vehicle Systems association (UK) (www.uavs.org) – the oldest non-profit trade association devoted to the unmanned aerial systems community.

UAV Websites

Air Force UAV Simulation:
The web site
http://www.airforce.com/interactive/uav/index.html
lets you see what it's like to fly a UAV. The series of
videos includes a short introduction, a training
manual, and a simulated search and destroy mission
piloting an MQ-9 Reaper UAV.

AUVSI (Association for Unmanned Vehicle Systems International):
AUVSI (http://www.auvsi.org) is dedicated to
promoting unmanned systems and related
technologies. AUVSI is a non-profit organization and
includes over 6,000 members representing 2200
companies in 55 countries.

sUAS News:
http://www.suasnews.com is a primary news source
for information on smaller UAVs.

AOL UAV News:
http://www.aolnews.com/tag/uav/ offers news,
photos and background information on UAVs.

UAV News:
http://www.spacewar.com/uav.html provides links
to current news stories involving UAVs.

Military Industry News on UAVs:
http://military.einnews.com/news/uavs provides
news on developments in the UAV industry.

Remotely Piloted Aircraft Forum:
http://www.flyingsquadron.com/forums/index.php?/forum/22-remotely-piloted-aircraft-rparpvuasuav/ is a forum for current news and developments involving UAVs.

UAV Forum:
http://www.rcgroups.com/uav-unmanned-aerial-vehicles-238/ offers forum posts regarding radio-controlled recreational UAVs and UAVs in general.

DIY Drones:
DIY Drones (www.diydrones.com) helps UAV hobbyists construct their own unmanned flying vehicles. There are forums, blogs, helpful articles and an online store where you can buy complete kits or just the parts to build your own UAV. DIY Drones offers a wide range of UAV kits like the "BlimpDuino" (priced around $90), a flying balloon with an Arduino board that controls the balloon's propulsion and navigation. Other UAVs for sale include airplanes, helicopters, quadcopters and blimps.

Games/Videos/Books

Are there UAV video games?
There are a few UAV games out there. For example:
- UAV Predator ($28.53 at amazon.com) (Windows Vista/2000/XP) – Fly a Predator from a ground station.
- UAV Fighter (iPhone) (Apple App Store)
- The web site http://www.airforce.com/interactive/uav/index.html lets you fly a simulated UAV mission.

Any books featuring UAVs?
There are a number of non-fiction books about UAVs and Dale Brown has authored several fiction books in the "Dale Brown's Dreamland" series that feature futuristic UAVs called "Flighthawks".

Can playing video games help me fly UAVs?
Researchers at the Massachusetts Institute of Technology (MIT) and other groups are trying to determine if there's a link between video game expertise and superior UAV controllers. So far no direct correlation has been found, but Senior Chief Aviation Electronics Technician Stephen Diets who is training to be an operator of the MG-8 Fire Scout feels that his background playing games on the Atari and Playstation is helpful. According to Diets, the things you have to keep track of in some video games (maps, compasses, ammo loads) are the same type of things you have to deal with in controlling a military UAV. Games like "Tom Clancy's Ghost Recon – Advanced Warfighter" even let

players fly a UAV over a digital battlefield to detect threats. Just be aware – actually flying a military UAV isn't nearly as exciting as a video game. It's serious business and requires the ability to concentrate and stay alert over a long period of time.

Is the U.S. Military making use of Video game technology for UAV training?
Realistic video game simulations are rapidly becoming a part of military UAV training courses. For example the Army's UAV training courses at Fort Huachuca now let trainees spend a considerable amount of time flying simulated drones on virtual missions over areas similar to that in Iraq and Afghanistan. Each simulator uses digitized satellite imagery from actual locations in Iraq and Afganistan to provide realistic terrain and instructors can input simulated targets or items of interest. Trainees do mission planning based on intelligence provided to them, make out a list of "targets of opportunity" to check out. Then they "fly" the mission, attempting to find the targets of opportunity (not necessarily targets to be fired on) and anything else pertinent. Once the mission is over, instructors review the results and critique the trainees on their performance.

Also, a company named L-3 Link has created Predator and Reaper simulations (used by the Air Force and Air National Guard) that mimic different times of day, various weather conditions, and include life-like characters and realistic scenarios. Ultimately the military would like to have simulations that trainees could take home with them, run the games on their own PC and study various situations and outcomes – a method of providing realistic "homework" in other words.   The center also trains mechanical and electronic technicians to provide maintenance for the UAVs.

Appendix A - Terminology

AHRS – Altitude Heading Reference System
ALR – Automatic Launch and Recovery
ATARS - Advanced Tactical Airborne Reconnaissance System
ACC – Air Combat Command
AFCS – Automatic Flight Control System
AGL – Above Ground Level
ASIP – Airborne Signals Intelligence Payload
ATO – Air Tasking Order
AUVSI – Association for Unmanned Vehicle Systems International
BLOS – Beyond Line Of Sight (communications)
CAS – Close Air Support
DARPA – Defense Advanced Research Projects Agency
EW – Electronic Warfare
FAA – Federal Aviation Administration
FAR – Federal Aviation Regulation
FCS – Flight Control System
FLIR – Forward Looking Infrared
GCS – Ground Control Station
GPS – Global Positioning System
HALE – High Altitude Long Endurance
IFR – Instrument Flight Rules
INS – Inertial Navigation System
ISR – Intelligence, Surveillance and Reconnissance
JTAC – Joint Terminal Attack Controller
LALE – Low Altitude Long Endurance
LOS – Line Of Sight (communications)
LRE – Launch and Recovery Element
MALE – Medium Altitude Long Endurance

MARS – Mid-Air Retrieval System

NAS – National Air Space

NOTAM – Notices to Airmen

MCE – Mission Control Element

OIF – Operation Iraqi Freedom

OEF – Operation Enduring Freedom (Afghanistan)

RATO – Rocket Assisted Take Off

ROA – Remotely Operated Aircraft

RPAV – Remotely Piloted Aerial Vehicle

RPV – Remotely Piloted Vehicle

R/C (or RC) – Radio Controlled

SLAR – Side-Looking Airborne Radar

STUAS – Small Tactical Unmanned Aircraft Systems

sUAS – Small Unmanned Aircraft Systems

UAS – Unmanned Aerial System

UAV – Unmanned Aerial Vehicle

UCAV – Unmanned Combat Air Vehicle

VFR – Visual Flight Rules

VLOS – Visual Line Of Sight

WAAS – Wide Area Augmentation System

Appendix B – Types of UAVs

Note: The naming scheme for UAVs works as follows: the letter "R" stands for Reconnaissance, the letter "M" stands for Multi-role, "Q" stands for an unmanned aircraft system, and the addition of a letter "A", "B", etc. at the end of the name indicates a modified or enhanced version of the original UAV.

MILITARY:
- **RQ-11B Raven** - The Army has more than 7,000 hand-launched Raven UAVs in service, making it the most prevalent UAV around (and the Raven is now in use by the Marine Corps as well). The Raven is 3 feet long and weighs 4.2 pounds and is typically fitted with an electronically stabilized color video camera or an infrared video camera for night missions, providing ground troops with an overview of their immediate surroundings.. Effective range for the RQ-11B is about six miles.
- **Wasp III** - Used by U.S. Special Forces units, the Wasp III is electric-powered and extremely quiet. It is hand-launched, weighs just one pound, flies 20 to 40 miles per hour at altitudes up to 500 feet and can be programmed to fly a completely autonomous mission.
- **Desert Hawk** - Used by American and British troops in Afghanistan, the Desert Hawk is hand-launched and follows pre-programmed coordinates to give troops an "over-the-hill" view day or night up to six miles away.

Extremely durable, the Desert Hawk is built of expanded polypropylene and fitted with Kevlar skids.

- **T-Hawk** - Used by U.S. Army infantry in Iraq. The VTOL T-Hawk can fly at altitudes as high as 10,000 feet for up to 45 minutes.
- **Scan Eagle** - Used by U.S. Marines, the Scan Eagle is 4 feet long and weighs about 40 pounds. It's powered by a gasoline engine and launches from a catapult.
- **RQ-7 Shadow** - Used in Iraq and Afghanistan to provide tactical surveillance. The Shadow launches from a catapult and can stay aloft for five to six hours, flying at altitudes up to 14,000 feet. It's about 11 feet long, weighs around 375 pounds and is equipped with an infrared illuminator, allowing it to provide laser targeting for laser-guided missiles and bombs. Note: The Shadow can also deliver a 20 lb. "Quick-Meds" canister to front-line troops.
- **MQ-1 Predator** - Used extensively by the United States in Iraq, Afghanistan and Pakistan. The Predator is an armed reconnaissance UAV which can carry two Hellfire air-to-ground missiles. It has a range of about 400 nautical miles and can stay airborne for up to 24 hours.
- **FINDER** (Flight Inserted Detection Expendable for Reconnaissance) - The FINDER is actually a UAV which can be launched from another UAV (a Predator), to get a closer view of a target. It is just over 5 feet in length, weighs about 60 pounds and can be flown using the Predator's

controls. The FINDER launches like a rocket and then its wings unfold for flight.

- **MQ-1C Gray Eagle** - An upgrade to the MQ-1 Predator developed for the U.S. Army, the Gray Eagle is an Extended Range Multi-Purpose (ERMP) UAV powered by a diesel engine that burns jet fuel. It is capable of operating for up to 36 hours at a time with a maximum altitude of 25,000 feet and an operating range of 200 nautical miles. The MQ-1C's nose houses a Synthetic Aperture Radar/Ground Moving Target Indicator (SAR-GMTI) system and an AN/AAS-52 Multi-spectral Targeting System (MTS). It can carry a payload of 800 pounds and can be armed with AGM-114 Hellfire missiles and GBU-44/B Viper Strike laser guided bombs.
- **MQ-5 Hunter** - Flown by the U.S. Army in Iraq and Afghanistan. The Hunter has been around for a while and was recently refitted as the MQ variant to carry Viper Strike munitions. It has a 34 foot wingspan, can fly 18 hours at altitudes up to 18,000 feet and can be flown from the same ground control station as the Army's version of the MQ-1 Predator.
- **MQ-8B Fire Scout** – The Fire Scout is an unmanned autonomous helicopter designed to provide reconnaissance and targeting support. The MQ-8B is fitted with a four-blade main rotor to reduce noise and improve lift. It can carry Hellfire missiles, Viper Strike laser-guided weapons, and the "Advanced Precision Kill Weapon System (APKWS), a laser-guided

70 millimeter folding-fin rocket. The MQ-8B is used by both the Army and the Navy.

- **MQ-9 Reaper** - The Reaper is a bigger, beefier cousin of the MQ-1 Predator and is also in use in Iraq, Afghanistan and Pakistan. Twice the size of a Predator, the Reaper is jet-powered and can carry 3,000 pounds of ordinance, including Hellfire missiles and the GBU-12 laser-guided bomb. After launch the Reaper is normally flown via satellite link by crews in the United States.
- **RQ-4 Global Hawk** - Used by U.S. forces in Iraq, Afghanistan and Pakistan, the Global Hawk is 44 feet long, weighs about 8500 pounds and can stay aloft up to 36 hours. It can fly at altitudes up to 65,000 feet and provide coverage of over 40,000 nautical square miles. The RQ-4 is powered by an Allison Rolls-Royce turbofan engine and carries a payload of 2,000 pounds. The Global Hawk features a full suite of infrared, electro-optical and radar sensors and can operate autonomously day or night in all types of weather.
- **RQ-170 Sentinel** - The highly-classified Sentinel flies sorties over Afghanistan (and possibly elsewhere), controlled by satellite link from Creech AFB in Nevada. It's an offspring of Lockheed Martin's Skunk Works program, but very little else is known about it. It is rumored to have been used in trying to track down Osama Bin Laden.
- **CU-170 Heron** – A Medium Altitude Long Endurance (MALE) UAV developed by Israel,

the Heron can stay airborne for over 50 hours at altitudes up to 35,000 feet. The Heron is capable of either pre-programmed flight or ground controlled flight and has fully automatic launch and recovery (ALR) and all-weather capabilities. It can carry a variety of sensors and can also be used for target acquisition and artillery adjustment.

- **Hermes 450/Watchkeeper** - Used by Israel and by British troops in Iraq and Afghanistan, Hermes can stay aloft for about 20 hours providing real-time surveillance to commanders in the field.

EXPERIMENTAL:

- **Phantom Ray** - The Phantom Ray is a jet-powered stealth UAV, shaped like a flying wing. It is currently being used to test advanced UAV capabilities like radar jamming, aerial refueling, and air-to-air missile combat. The Phantom Ray has a ceiling of about 40,000 feet and an estimated maximum speed of a little over 600 miles per hour.
- **Demon** - A wing-shaped, highly maneuverable UAV being developed by BAE Systems in England.
- **Vulture** - Vulture is Lockheed Martin's design for DARPA's Vulture program. It is designed to stay aloft for up to five years to provide continuous monitoring of some area for long periods of time. Vulture is equipped with a suite of day-and-night cameras that can scan a 600-mile area, sending imagery and other data back to handlers on the ground. A semiflexible

structure allows the Vulture to bend instead of breaking when high winds cause it's lengthy frame to oscillate violently.

- **Global Observer** – AeroVironment is developing the Global Observer, a hydrogen-powered high-altitude long-endurance (HALE) UAV. Global Observer will be capable of flights of at least 5 days in duration at altitudes up to 65,000 feet. Note: A GO-1 Global Observer crashed at Edwards AFB on April 1, 2011 while undergoing testing and future funding for the program could be a problem.

CIVILIAN:
- **Bell Eagle Eye** – Bell's Eagle Eye is a tilt-rotor UAV similar in design to the V-22 Osprey. Eagle Eye has been around since 1998 in one version or another, and so far has failed to generate much market interest although the U.S. Coast Guard has tested it for use as a ship-based search-and-rescue platform.
- **Yamaha RMAX** – The RMAX is a 12-foot long unmanned helicopter and is in use by the city of Tucson, Arizona as a platform to spray wetlands for mosquito control.
- **SkySeer** – An electric-powered, short-range, hand-launched UAV is being experimented with as a means of crime prevention. A complete SkySeer system retails for $25,000 to $30,000 including the UAV, its electronics, video surveillance equipment, ground station and computer inferfaces.
- **Aerosonde** – A small UAV designed to collect weather data (temperature, air pressure, wind

and humidity) over oceans and remote areas, the Aerosonde was the first UAV to cross the Atlantic. Launched from the roof rack of a moving car, the Aerosonde flew from Newfoundland, Canada to an island off the coast of Scotland in less than 27 hours, burning about 1 ½ gallons of gasoline. Aerosondes were also the first UAVs to fly into the eye of a tropical cyclone. The Aerosonde is about 5 ½ feet long with a wingspan of 9 ½ feet, a maximum speed of 90 miles per hour and a maximum altitude of 15,000 feet.

Appendix C – R/C Ground School

A good first step to learning to fly UAVs is to learn how to fly a radio-controlled (R/C) airplane. Here's a quick "ground school" on flying R/C models.

R/C airplanes are controlled by a radio signals transmitted from a hand-held transmitter to a receiver in the airplane. Older models used transmitters in the megahertz range but 2.4 gigahertz (GHz) radios are becoming standard since they are far less susceptible to interference.

Whatever type of transmitter is used the signals are picked up by the receiver and converted into physical movement through the use of servos inside the airplane that are linked to the model's control surfaces.

The control surfaces are used to change the plane's position or altitude or to correct for unintentional changes. When an airplane changes position in flight it rotates around one or more of three axes (imaginary lines that pass through the aircraft's center of gravity). Motion around the lateral axis (motion up and down) is referred to as **pitching**. Motion around the longitudinal axis is called **rolling**, and motion around the vertical axis (motion to one side or the other) is called **yawing**.

The primary control surfaces are the **rudder, elevator and ailerons**. The most basic R/C models (single channel models) will only control the rudder (the hinged portion of the vertical stabilizer on the tail of the airplane. The **rudder** controls the **yaw –** when the rudder moves to the left the plane turns left and vice versa.

The **elevators** are the hinged sections of the horizontal stabilizer on the tail of the airplane and they control the **pitch attitude** of the plane – whether the nose of the plane is pointing up, down or level. Moving the elevators upward pushes the airplane's tail downward and the nose up (increasing the angle between the plane and the streaming air – the **angle of attack**) and the plane starts to climb. Moving the elevators downward causes the nose to come down and the airplane will start to dive.

The **ailerons** are the hinged sections of the trailing edge of each wing. Ailerons always move in the opposite direction to each other – when the aileron on the left wing goes up, the aileron on the right wing goes down and vice versa. Ailerons control the **roll** of the airplane about its longitudinal axis. Moving the right aileron up decreases the lift on the right wing and causes the left aileron to move downward (increasing the lift on the left wing) and the plane rolls to the right. Moving the right aileron down (and the left aileron up) rolls the aircraft to the left. A side effect of using the ailerons is **adverse yaw** – a yawing movement in the opposite direction to the roll caused primarily by the change in **drag** on the left and right wings. Adverse yaw is compensated for by use of the **rudder**.

Some RC airplanes also have **flaps**, which are also hinged sections on the trailing edge of the wing. If the airplane happens to have flaps, they will be separate from the ailerons and located on the inside portion of the trailing edge of the wing. The ailerons will be on the outside portion of the trailing edge. Flaps work together – they go up or down at the same time and serve to increase or decrease the airflow over the wings, increasing or decreasing lift. Flaps are used primarily in landing; lowering the flaps increases the lift under the wing and helps keep the airplane in the air as it slows down to land.

Because of their design some aircraft don't have all of these separate control surfaces – they use control surface **mixing**. For example, the elevators and ailerons may be combined or "mixed" and are called **elevons**. Elevons look like elevators but they can move together (like elevators) or separately (like ailerons). Similarly **flaperons** mix the actions of flaps with ailerons, with one pair of control surfaces along the trailing edge of the wing doing the job of both aileron control and flap control.

Every controllable operation on an R/C plane is referred to as a **channel**. A single channel model can only control one type of action – motor (on or off), rudder, ailerons, etc. A 4 channel model might be able to control the rudder, elevator, ailerons and the plane's motor.

However many channels your R/C plane has, the first step in flying your plane is to go through a pre-flight checklist:
- Make sure all screws or other fasteners are tight

- Check to make sure all control surfaces (rudder, ailerons, etc.) move freely and respond properly to commands from the transmitter
- make sure the receiver, battery pack or other electronic gear is fastened securely inside the plane
- If the plane is powered by an internal combustion engine check the fuel tank and tubing – if the plane is battery powered make sure the batteries are fully charged
- Check to make sure no wires or other items are in position to interfere with the servos
- Make sure the receiver is receiving commands from the transmitter

Once you've gone through the pre-flight checklist it's time to get your plane airborne.

To launch the plane, make sure it's facing directly into the wind. Always take off into the wind because that maximizes the airflow over the wings and helps the aircraft get airborne sooner. Slowly increase power to full, letting the airplane accelerate over the ground. You will probably need to use the rudder to keep the plane moving in a straight line – don't let it veer off to the left or right.

The plane should take off fairly quickly – if it's struggling to get airborne, try giving it a small amount of up elevator to lift the nose. Don't apply full elevator as the plane starts its takeoff run – the plane will likely take off, climb at too steep an angle, stall out and crash. Once the plane is airborne remember to execute a turn before it goes out of range of the transmitter.

Hand launching works basically the same as letting the plane take off on its own. Hold the airplane about head-high and bring the motor to maximum power, then take a step forward and give the plane a firm push, keeping the aircraft level or very slightly nose-up.

Once the model is airborne you need to **"trim"** the airplane. Use the trim controls on the transmitter to correct any tendency of the airplane to turn, bank, climb or dive on its own.

To land the plane, be sure to land by flying a **"downwind leg"** (with the wind) and then turning into the wind. Landing into the wind increases the airflow over the wings and helps keep the plane from stalling out as it slows down. As you turn into the wind start to slow the plane down and use the rudder to keep it in a straight line. As the airplane nears the ground reduce power to the motor completely and apply a little up elevator to slow the plane's speed and rate of descent further, until it touches down. This maneuver is called **"flaring"** and the timing is critical – flare too soon and the plane will stall and crash, too late and you're going to have a very rough landing.

Basic maneuvers include:
- Roll – to execute a roll start by flying straight and level on at least half throttle. Apply a small amount of up elevator and left or right

aileron (depending on which direction you want to roll). As the plane starts to roll inverted, stop applying up elevator, then once the plane is fully inverted, use a very small amount of down elevator to maintain altitude and let the airplane complete the roll.

- Inside Loop – Start by flying straight and level into the wind, then open the throttle and apply full power while pulling back gently on the elevator stick to start a climb. Let the plane climb until it starts to roll over onto its back, then close the throttle and keep holding the elevator stick back. As the plane completes the loop move the elevator stick back to neutral to resume level flight.
- Outside Loop – Start the loop by rolling the plane inverted (upside down) and using down elevator to begin climbing. Reduce power at the top of the loop and as the plane completes the loop roll the airplane right side up again.

Appendix D – Flight Instruments

Most aircraft have certain standard flight instruments (although in the case of UAVs these instruments will appear as readouts on a workstation screen rather than as actual gauges):

- **Altimeter** – Shows the aircraft's altitude above sea-level by measuring the pressure difference between a stack of aneroid capsules inside the altimeter and the atmospheric pressure outside the aircraft.
- **Airspeed Indicator** – Shows the aircraft's speed (usually in knots) relative to the surrounding air. The airspeed indicator has to be corrected for the density of the air, which varies with temperature, altitude and humidity, to get the true airspeed and for wind conditions to get the speed over the ground.
- **Attitude Indicator (Artificial Horizon)** – Shows the aircraft's attitude (position in space) relative to the horizon. The attitude indicator lets the pilot know if the wings are level and if the aircraft's nose is pointing above or below the horizon. This instrument is particularly important in bad visibility or for instrument flight.
- **Magnetic Compass** – Shows the aircraft's heading relative to magnetic north. Because turning, climbing, descending or accelerating can cause the magnetic compass to give confusing indications, the **Heading Indicator**

is also used to determine the aircraft's true heading.

- **Heading Indicator (Directional Gyro or Gyrocompass)** – Shows the aircraft's heading relative to geographical north. The Heading Indicator is basically a spinning gyroscope and as such is subject to drift errors (precession) which are normally corrected periodically by calibrating the heading indicator to the magnetic compass. In more advanced aircraft the heading indicator has been replaced by a **Horizontal Situation Indicator (HSI)**.
- **Turn Indicator** – Shows the direction of turn and the rate of turn.
- **Vertical Speed Indicator** – The VSI displays changing air pressure as a rate of climb or descent.

In addition to these instruments the controls in a UAS will also include items such as autopilot status, GPS display, data link readouts, digital maps and waypoint displays, telemetry data and video images.

Alphabetical Index

Notes: